DEFENDING NATURE

HOW THE US MILITARY PROTECTS THREATENED AND ENDANGERED SPECIES

SNEED B. COLLARD III

M Millbrook Press / Minneapolis

This book is dedicated to both the military and civilian men and women and their families who have kept, and continue to keep, our nation safe.

A special acknowledgment to biologist Bruce Hagedorn, whose vision guided so much of the work documented in this book.

Text copyright © 2025 by Sneed B. Collard III

All rights reserved. International copyright secured. No part of this book may be reproduced, stored in a retrieval system, or transmitted in any form or by any means—electronic, mechanical, photocopying, recording, or otherwise—without the prior written permission of Lerner Publishing Group, Inc., except for the inclusion of brief quotations in an acknowledged review.

Millbrook Press™
An imprint of Lerner Publishing Group, Inc.
241 First Avenue North
Minneapolis, MN 55401 USA

For reading levels and more information, look up this title at www.lernerbooks.com.

Illustrations on pages 8, 11, 13, 19, 30, and 40 by Laura K. Westlund.

Designed by Kimberly Morales.
Main body text set in Cisalpin LT Std. Typeface provided by Linotype AG.

Library of Congress Cataloging-in-Publication Data

Names: Collard, Sneed B., author.
Title: Defending nature : how the US military protects threatened and endangered species / by Sneed B. Collard III.
Description: Minneapolis : Millbrook Press, [2024] | Includes bibliographical references and index. | Audience: Ages 9–14 | Audience: Grades 4–6 | Summary: "On Eglin Air Force Base, scientists and soldiers work together to protect endangered red-cockaded woodpeckers, gopher tortoises, and reticulated flatwoods salamanders. Discover the challenges endangered species face and how the military works to protect them" —Provided by publisher.
Identifiers: LCCN 2023021110 (print) | LCCN 2023021111 (ebook) | ISBN 9781728493749 (library binding) | ISBN 9798765619186 (epub)
Subjects: LCSH: Endangered species—Conservation—Florida—Pensacola—Juvenile literature. | Military bases—Florida—Pensacola—History—Juvenile literature. | BISAC: JUVENILE NONFICTION / Animals / Endangered
Classification: LCC QL84.22.F6 C65 2024 (print) | LCC QL84.22.F6 (ebook) | DDC 591.6809759/99—dc23/eng/20230713

LC record available at https://lccn.loc.gov/2023021110
LC ebook record available at https://lccn.loc.gov/2023021111

Manufactured in the United States of America
1-53603-51384-3/28/2024

CONTENTS

- **MISSION CANCELED** / / / / / / / / / / / / / / / 4
- **CHAPTER ONE**
  ACCIDENTAL REFUGES / / / / / / / / / / / / / 6
- **CHAPTER TWO**
  FIREPOWER AND FORESTS / / / / / / / / / / / 10
- **CHAPTER THREE**
  WOODPECKER WARRIORS / / / / / / / / / / / 18
- **CHAPTER FOUR**
  ZOMBIE TORTOISES / / / / / / / / / / / / / 26
- **CHAPTER FIVE**
  SALAMANDER SOLUTIONS / / / / / / / / / / 36
- **CHAPTER SIX**
  THE BIGGER MISSION / / / / / / / / / / / / 44

AUTHOR'S NOTE	50
GLOSSARY	51
SOURCE NOTES	52
FOR FURTHER INVESTIGATION	53
SELECTED BIBLIOGRAPHY	54
INDEX	56

MISSION CANCELED

In 1989 mission planners at Eglin Air Force Base near Pensacola, Florida, were preparing for a munitions test using explosive materials. One of Eglin's main roles was to test weapons and aircraft systems, and with almost half a million acres (200,000 ha) of land at its disposal, the base had plenty of room for these activities. Air force officials had spent months getting ready for the test, including sharing their plans with the US Fish and Wildlife Service (USFWS). This last step was required because Eglin was home to a number of endangered species, including a small bird called the red-cockaded woodpecker. These species are protected by law, so the military has to consult with the USFWS before doing anything that might pose a risk to the endangered plants and animals on their bases.

Eglin Air Force Base includes many diverse land and water habitats that protect dozens of endangered and threatened species.

The USFWS read over the test plans and asked air force officials a few questions:

- How many red-cockaded woodpeckers, or RCWs, lived at Eglin?
- Where did the birds live on the base?
- Would the planned test harm or kill any woodpeckers?

The air force gave the same answer to all these questions: we don't know.

As a result, the USFWS issued Eglin a jeopardy biological opinion. This document stated that the planned test might threaten the woodpeckers on the base, and that the USFWS could not approve the operation. Eglin still could go ahead with the test, but the base could be sued or taken to court if any woodpeckers were harmed.

The jeopardy opinion sent shock waves through officials at Eglin and rippled across the rest of the US Department of Defense. No one recorded the exact conversation that occurred at Eglin following the decision, but it went something like this:

"Can they do this?" fumed Eglin's commander at the time. "Can they threaten a military operation because of some *birds*?"

"Yes, they can," air force lawyers told him.

Reluctantly, Eglin canceled the test, and the base commander met with the base's chief of the Natural Resources Branch. "I do not want this to ever, *ever* happen again," the commander told him. "Do what it takes to make sure that it doesn't."

From that point on, conservation at Eglin took a radical turn. Instead of viewing environmental protection as a barrier or an inconvenience, Eglin's leadership began to embrace conservation as an essential part of its operations—one that benefits nature, Eglin's military mission, and the United States as a whole.

The endangered red-cockaded woodpecker helped spark new attitudes toward conservation at Eglin Air Force Base and across the Department of Defense.

» CHAPTER ONE «
ACCIDENTAL REFUGES

> Military bases such as Marine Corps Base Camp Pendleton in California often serve as islands of biodiversity amid growing towns and cities.

Eglin Air Force Base is one of more than 1,000 military bases operated by the United States Department of Defense, or DoD. These include 450 to 500 bases on US soil and 750 to 800 bases overseas. Some bases are small, staffed with only a dozen or so people. Others stretch across hundreds of thousands of acres and are home to tens of thousands of military personnel. The bases hold all six branches of the DoD: the army, navy, air force, marines, coast guard, and space force.

Military and civilian personnel at these bases perform countless activities to keep the United States safe and protect its interests around the world. They house and train combat soldiers, maintain ships and aircraft, keep watch over enemy nations, provide support to allies, and respond to both security and humanitarian crises. In recent decades, military bases have assumed another essential role: protecting animal and plant species.

It's a role that happened almost entirely by accident.

ISLANDS OF LIFE

Most US military bases were established during or before World War II (1939–1945). Back then, it was still fairly easy to set aside large tracts of land with few or no people on them. Installations such as Fort Bragg in North Carolina, Camp Pendleton in California, and Fort Cavazos (formerly Fort Hood) in Texas carved out hundreds of square miles of land at a time when the US human population was only about a third of what it is today. Many of these bases are in locations that would become very desirable places to live. As the US population boomed after World War II, cities and towns sprouted up around these installations, destroying habitat for animal and plant species. The only places some species survived were on the military bases themselves.

Justin Johnson is Eglin's supervisory wildlife biologist and oversees much of the conservation work on the base. Johnson explains, "The bases weren't intended for [conservation], but that's the way things worked out. Today, almost five hundred threatened and endangered species live on military installations that were not set aside for biodiversity and conservation." As the land around the bases became more densely populated, he continues, it created islands of biodiversity inside of these protected military areas, where a lot of species can still be found.

To put this in perspective, DoD lands hold more threatened and endangered species than the United States' National Parks System. Fifty-five threatened and endangered species live *only* on military installations. So do dozens of at-risk species—those that might soon become

CIVILIANS IN THE MILITARY

We usually think of all people working for the military as having a rank such as private, sergeant, captain, or general. About one-quarter of DoD employees, though, are civilian professionals such as Justin Johnson. They do not hold ranks but are hired by the Department of Defense to do specific jobs. One of those jobs is conservation, and Eglin employs a team of about fifty civilian biologists, foresters, and other personnel to carry out that work.

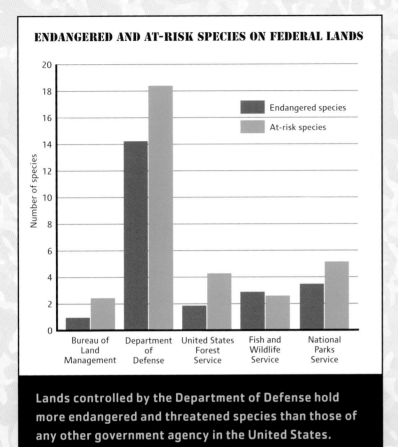

ENDANGERED AND AT-RISK SPECIES ON FEDERAL LANDS

Lands controlled by the Department of Defense hold more endangered and threatened species than those of any other government agency in the United States.

One of the first animals protected by the Endangered Species Act was the Mississippi sandhill crane, a subspecies of sandhill crane.

threatened or endangered. Looking out for these plants and animals falls squarely on the Department of Defense.

SIKES AND THE ENDANGERED SPECIES ACT

The US military's duty to protect nature has its origins in two landmark laws. Most people have heard of the Endangered Species Act, signed by President Richard Nixon on December 28, 1973. This sweeping legislation provides protections for threatened and endangered species and requires that the United States create plans for their recovery. It grants two agencies, USFWS and the National Oceanic and Atmospheric Administration (NOAA), broad powers to list species as threatened and endangered and to oversee their protection. The USFWS and NOAA work with many other international, federal, and state agencies—including the DoD.

The lesser-known Sikes Act of 1960 is just as important as the Endangered Species Act. The Sikes Act was developed by Robert Sikes, one of the most powerful politicians in Washington, DC, from the 1940s to the 1970s. He was also the congressman who represented the Florida district where Eglin Air Force Base sits.

The Sikes Act has two major objectives. It requires the DoD to work with state and local governments to create hunting, fishing, and other recreational opportunities on military lands whenever these activities don't interfere with military operations. It also requires that the military work to conserve, protect, and rehabilitate species and ecosystems on DoD bases.

Together, the Endangered Species Act and the Sikes Act provide a clear mandate to the US military to protect nature on the lands it oversees. Nonetheless, the DoD placed a low priority on this obligation until the early 1990s. Around the time that Eglin received its jeopardy biological opinion for the red-cockaded woodpecker, military attitudes began to change. As more people became aware of the environmental importance of military bases, the DoD began investing more resources into conservation. Nowhere has this been more apparent than at Eglin Air Force Base.

Robert Sikes is credited with helping bring fourteen military bases to Florida and creating a legacy of protecting nature on military installations.

MILITARY INSTALLATIONS WITH LARGE NUMBERS OF THREATENED AND ENDANGERED SPECIES

MILITARY BRANCH	INSTALLATION	NUMBER OF SPECIES
Army	Schofield Barracks Military Reservation, HI	58
Army	Makua Military Reservation, HI	46
Navy	Joint Base Pearl Harbor-Hickam, HI	41
Army	Kawailoa Training Area, HI	33
Navy	Joint Base Marianas, Guam	32
Army	Pohakuloa Training Area, HI	26
Air Force	Homestead Air Reserve Base, FL	25
Air Force	Vandenberg Space Force Base (formerly Vandenberg Air Force Base), CA	23
Air Force	Cape Canaveral Space Force Station (formerly Cape Canaveral Air Force Station), FL	21
Air Force	Eglin Air Force Base, FL	19

This table shows some of the military bases with the largest numbers of threatened and endangered species that are federally protected under the Endangered Species Act.

CHAPTER TWO
FIREPOWER AND FORESTS

During its long history, Eglin has served as a crucial testing and operations base for many essential military aircraft, including the United States' longest-serving bomber, the B-52.

Like many other US military bases, Eglin Air Force Base has its origins in World War II. A small bombing and gunnery range operated in the area during the 1930s. But it wasn't until the outbreak of fighting in Europe that the US War Department took over 384,000 acres (155,000 ha) of the Florida Panhandle's Choctawhatchee National Forest next to the Gulf of Mexico. Eglin's official website states, "In 1941, the Air Corps Proving Ground was activated, and Eglin became the site for gunnery training for Army Air Forces fighter pilots, as well as a major testing center for aircraft, equipment, and tactics."

During World War II, Eglin played a key role in developing new aircraft and weapons systems used to defeat Japan and Germany. After Japan's attack on Pearl Harbor, Hawaii, on December 7, 1941, Lieutenant Colonel Jimmy Doolittle's B-25 aircraft crews used Eglin to practice their historic bombing raid on Tokyo. Later, US Army Air Forces pilots developed strategies to destroy missile installations the Germans used to launch rockets at England, one of the United States' allies.

Following the war, Eglin's importance to the US military grew. Its large size and isolation from major cities made it ideal for developing guided missiles, laser-guided bombs, anti-tank weapons, and air-to-air fighting tactics that have been used in Korea, Vietnam, Iraq, Afghanistan, and other military conflicts of the past eighty years. The base also provides training grounds both for general combat troops and special forces such as the army's 7th Special Forces Group (Airborne) and the 6th Ranger Training Battalion.

Eglin serves as headquarters for the 96th Test Wing, one of the US military's major command units responsible for developing and testing advanced weapons systems. In addition to using the base's 464,000 acres (188,000 ha) of land, the military conducts air training and testing operations across 120,000 square miles (310,000 sq. km) of the Gulf of Mexico. The entire area is critical not only to the military's mission but also to the incredible variety of plants and animals living there.

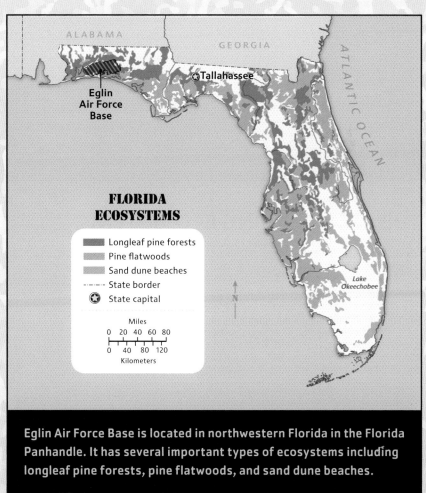

Eglin Air Force Base is located in northwestern Florida in the Florida Panhandle. It has several important types of ecosystems including longleaf pine forests, pine flatwoods, and sand dune beaches.

THE F-35: SOUND AND FURY

While dozens of military units call Eglin home, none capture the spotlight like its F-35 Lightning II combat squadrons. The base serves as one of the major training centers for F-35 pilots, so it's a rare day when Eglin's biologists don't hear the roar of the jets overhead. Many experts consider the F-35 the most advanced fighter aircraft ever built. Three variations of the plane operate out of Eglin: the standard F-35A flown by the air force; the short takeoff and vertical-landing F-35B piloted by the marines; and the navy's F-35C, which is designed for use on aircraft carriers.

Recognizable by the loud, ear-pounding roar of its engine, these fighter jets feature advanced electronic warfare capabilities, stealth technology, and a large fuel capacity that enables them to conduct long-range missions. They also can carry weapons for almost any combat mission. The DoD plans to purchase nearly twenty-five hundred of the planes through the year 2044. The US has also made these planes available to Britain, Israel, Japan, and other close allies.

The F-35 was designed to be upgraded to keep it as one of the world's best fighter aircraft for decades to come.

THE DISAPPEARING ECOSYSTEM

Eglin is home to at least thirty-four types of natural communities, or ecosystems, and it supports more than two hundred rare or officially protected species. But Eglin's dominant ecological feature—one that makes it especially important for animals and plants—is its extensive stands of longleaf pines.

Before European settlement in the Americas, 60 to 90 million acres (24 to 36 million ha) of longleaf pine forest stretched from present-day Virginia to Florida, and west to Texas. From the early 1800s to the end of the twentieth century, this forest habitat was whittled down to only about 3 million acres (1.2 million ha). Longleaf pine forests were logged for building ships and houses and converted to tree plantations for paper pulp and lumber. They were flooded by dams, turned into farms, and paved over to build American cities.

More than one thousand different plant species have been identified in longleaf pine forests. Approximately thirty-six mammal, ninety-six reptile, seventy-four amphibian, and one hundred bird species also live there. As the forests disappeared, so did its plants and animals. Nearly thirty plant and seven animal species that are listed as threatened or endangered by the USFWS

For the first several years of their lives, longleaf pines exist as a "grass stage." This stage can survive frequent fires while its root system develops enough to support growth of adult trees.

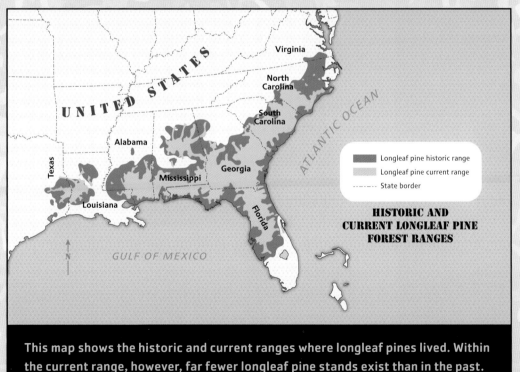

HISTORIC AND CURRENT LONGLEAF PINE FOREST RANGES

This map shows the historic and current ranges where longleaf pines lived. Within the current range, however, far fewer longleaf pine stands exist than in the past.

declined in population because of the destruction of longleaf pine forests. Eglin's pines fared better than most—and not just because they were locked up in a military base.

"One thing that saved Eglin's pines were our sandy soils," explains the base's chief of forest management, Al Sutsko. "A lot of the forest wasn't converted to agriculture like in the rest of the Southeast. The soils here just weren't fertile enough." Instead, the forest was mainly used for producing turpentine, a major industry a century ago. "Turpentine production, logging in small amounts, and some cattle grazing occurred here," said Sutsko, "but the forest wasn't cut down for growing cotton or peanuts, the traditional agricultural crops."

Bad luck for farmers proved to be good luck for the forest because more than 200,000 acres (81,000 ha) of longleaf pines survived on Eglin property. Scientists discovered, however, that even where the trees remained, many of the ecosystem's plants and animals dwindled or disappeared. For decades no one understood why, but the cause turned out to be simple.

Fire—or, rather, a lack of it.

TAPPING TURPENTINE

Turpentine is an oil that is used to make chemicals and as a solvent to dissolve or thin paint and other substances. Because Eglin's soils were not suited for agriculture, people during the early 1900s extracted turpentine and other products from the longleaf pines. They cut grooves into the trees and collected the sap that oozed out. This sap could be turned into turpentine, pitch, resin, and pine tar. These products—collectively called naval stores—had many uses, from waterproofing ships to manufacturing paints, insecticides, and other products.

Unfortunately, sap collection often killed adult trees, but the industry continued in the area until the forest became a military base. Since then, turpentine collection has been in decline. In some places, people still obtain turpentine by tapping live trees, but it is also produced as a by-product of paper manufacturing. Petrochemicals—chemicals made from petroleum, coal, and natural gas—have replaced its use in many manufacturing processes.

Turpentine collectors cut grooves at the base of trees where the resin collects. But these grooves left the trees vulnerable to disease.

THE FLAMMABLE FOREST

Much like ponderosa pine forests in the western United States, longleaf pine forests evolved alongside frequent fires. "Florida is very wet but is the lightning capital of the US, so lightning fires are a common occurrence here," explains wildland support module lead Brett Williams. "Native Americans also burned the landscape for various reasons including promoting certain plants and preventing severe wildfires. By looking at historical records and fire scars in tree rings, ecologists think that fires swept across this landscape every one to three years."

Frequent fires have probably burned through Florida for about the past ten thousand years, since the last ice age. They have had profound effects on the region's plants and animals. The longleaf pine itself may be the best example. Longleaf pines evolved to survive most fires, but they cannot grow when they are shaded out by other plants. Fires, however, kill competing tree species, allowing plenty of longleaf pines to survive.

A healthy longleaf pine forest looks different from most other forests. Instead of dense trees shading the undergrowth, frequent fires create open conditions almost like an African savanna. Longleaf pines are spaced far enough apart to ensure that plenty of sunlight reaches the ground, allowing a huge assortment of plant species to grow on the forest floor. The plants, in turn, support a large

Longleaf pine forests are naturally resilient to fire. As they grow, their lower branches are pruned by fires, which keeps mature trees' canopies above most flame heights.

variety of animals. Every two or three years, a new fire burns away excess plant material, resetting the system and keeping it productive and healthy.

For thousands of years, this longleaf-fire system worked wonderfully. It created ideal conditions for many different plants to thrive without getting shaded out or outcompeted by a few dominant species. Unfortunately, in the twentieth century, an unwelcome mammal showed up. His name? Smokey Bear.

THE PROBLEM OF FIRE PREVENTION

In 1910 an enormous fire known as the Big Burn torched 3 million acres (1.2 million ha) of northern Idaho and surrounding regions. The fire killed dozens of people and was seen as a major disaster by both the public and US Forest Service, which had been created in 1905. Several catastrophic fires had occurred before the Big Burn, but wildfires were generally considered a normal part of nature. The Big Burn ushered in an era when wildfires—even naturally occurring fires—were considered bad. The Forest Service started a series of educational campaigns to teach the public that all fires were dangerous and should be put out as soon as possible. The most famous campaign, which began in 1944, featured the Smokey Bear character along with Smokey's motto: "Only YOU can prevent forest fires."

Fire suppression policies lasted for decades—and with disastrous consequences. Because regular fires were no longer burning through many landscapes, huge quantities of deadwood and other fuels built up. This led to more intense, dangerous fires when they did occur. Trees that would have survived frequent, low-intensity burns were killed by the higher-intensity fires that broke out. Where fires were suppressed, both tall overstory trees and a few dominant understory shrubs and trees shaded out the many other plants that needed sunlight to grow. That, in turn, lowered the diversity of both plant and animal life.

Fire suppression took a particular toll on longleaf pine ecosystems. Without fire, other plant species moved in and grew larger, shading out almost everything living beneath them. Biodiversity in the forests plummeted. One animal species that suffered was a small black-and-white bird called the red-cockaded woodpecker.

PROTECTING EGLIN'S CULTURAL RESOURCES

In addition to protecting natural resources, Eglin is also responsible for protecting the base's cultural history through its Cultural Resources Management program. Native Americans lived on what is now Eglin for thousands of years before the arrival of Europeans. More than eight hundred Native American archaeological sites have been identified on the base, including the Bayou Park archaeological site identified in 1993. Like other DoD facilities, Eglin often hires private contractors to perform specific jobs. The base hired two companies to survey and excavate the Bayou Park site. There, workers uncovered more than half a million artifacts from a village that occupied the area between forty-five hundred and four thousand years ago.

Smokey Bear continues to warn the public about fire danger in many parts of the country.

»» CHAPTER THREE ««
WOODPECKER WARRIORS

Sap drips from the RCW cavity on this longleaf pine tree.

More than any other bird, the red-cockaded woodpecker—or RCW—is known for inhabiting longleaf pine forests. About twenty-two species of woodpeckers live in the United States, but none is like the RCW. Instead of living only in male-female pairs, it is one of the few woodpecker species that also lives in groups. These groups can consist of other helper birds that aid in caring for young and drilling holes, or cavities, in trees. Unlike most other species, RCWs avoid drilling into dead tree trunks and branches. Instead, they are the only North American woodpeckers that drill their cavities into a tree's living tissues.

As these birds drill into a living tree, the tree oozes sticky sap that runs down the trunk. Biologists believe this sap—loaded with irritating chemicals—deters snakes and other predators from climbing the tree, keeping the woodpeckers safe as they roost. The downside is that it takes an RCW group a *long* time to drill out a new nesting or roosting cavity. The birds peck a little bit, then wait for the area to dry or heal over so that fresh sap doesn't damage their feathers. While other woodpeckers can bang out a hole in a matter of days or weeks, it can take RCWs six months, a year, or longer to drill out a new cavity. These holes are used by more than just the woodpeckers that create them. Many other birds, mammals, and reptiles depend on unoccupied holes to live and rest in, and to raise their young. The woodpeckers and their holes are so important that scientists consider red-cockaded woodpeckers a keystone species—one that is essential to the health of the entire ecosystem.

Red-cockaded woodpeckers follow a similar life cycle as other birds, but the chicks and juveniles can be cared for by multiple birds instead of just their parents.

RCWs thrived in historic longleaf pine forests. The woodpeckers are counted as family groups instead of as individual birds, and scientists estimate that up to 1.6 *million* RCW groups once lived in the forests that stretched across the southeastern United States. When the longleaf pines began to disappear, however, so did the woodpeckers. By 1973, when the bird was officially protected under the Endangered Species Act, only a few thousand RCW groups survived.

BAD TO WORSE

Legal protection did not immediately improve the RCW situation. Scientists and land managers had not learned many basic facts about red-cockaded woodpecker biology, and destructive forestry practices continued to harm the birds. For instance, red-cockaded woodpeckers nest mostly in older pines that are infected with a fungus called red heart rot. This fungus softens the center of a tree—its heartwood—and makes it easier for the woodpeckers to drill into it. But the timber industry values *healthy* heartwood and in the 1960s, foresters began removing infected trees from forests, fearing the fungus would spread to uninfected trees. In doing so, they cut down the exact trees these woodpeckers nest and roost in.

Other disasters also impacted the longleaf pine forests. In 1989 Hurricane Hugo ripped through South Carolina's Francis Marion National Forest, home to the world's second-largest RCW population. The storm killed about half of the birds living there and destroyed 87 percent of their cavity trees. Almost everywhere, RCW populations continued to grow smaller and more widely separated. This made it difficult for young woodpeckers to disperse and colonize new territories.

Through the 1980s, the woodpeckers at Eglin Air Force Base fared no better. Forestry practices on the base focused on selling timber, not on restoring the health of the longleaf pine ecosystem. Even where there were plenty of older longleaf pines, many areas became too overgrown for RCWs to live there. Although the base held one of the largest populations of RCWs, no one knew how many woodpeckers remained or why they survived in some places and not in others. Eglin's biologists requested money and resources to monitor and manage RCW populations, but they were repeatedly turned down by base commanders.

The year 1989 would change all that.

This tree was infected with red heart rot. The middle of the trunk has softened and crumbled and much of the wood has turned reddish brown.

WOODPECKER WAKE-UP CALL

When Eglin received its jeopardy biological opinion, it proved to be a wake-up call for base leaders. They realized that if the base were to continue its extensive military operations, they would have to take a more active role in making sure that Eglin's biodiversity not only survived but *thrived*. Within months, Eglin's center commander approved money to conduct a thorough survey of the base's RCW population. The base also hired two new civilian staff members to help create an effective woodpecker management program. Eglin's Natural Resources Branch reached out to some of the world's top RCW experts to get their advice on the program. Their number one recommendation?

Fire.

"They were doing fire at Eglin before," explains Bruce Hagedorn, chief of Eglin's Natural Resources Branch, "but they were doing it mostly during the nongrowing season. It was not really having the same effects that were needed to manage for threatened and endangered species."

Historically, most fires at Eglin occurred during the growing season—the part of the year when plants are actively putting out new stems, leaves, and flowers. Burning during the *nongrowing* season created two problems. First, many of Eglin's plants need growing season fires to complete their full life cycles. Native wiregrass, for instance, only produces flowers when it burns during the growing season. The second problem is that most of Eglin's trees and other plants store the majority of their energy and nutrients underground. This is a good survival strategy, but it means that a fire during the nongrowing season doesn't touch these resources. Why? Because the plant isn't actively using them to create new leaves, shoots, and flowers. A growing season fire, on the other hand, burns up fresh leaves, shoots, and flowers. That literally consumes more

Wiregrass gets its name from the hairlike structures along its leaves.

of a plant's reserves, making it less able to compete with plants around it. By switching to burning during the growing season, Eglin's fire managers could keep fast-growing plants in check—and improve the overall health of the ecosystem.

Effective fires were especially important for the red-cockaded woodpeckers because tall understory plants allow predators to reach their nest holes, which drives woodpeckers away. The places in Eglin where the RCWs held on the best were near bombing ranges. The explosions of bombs and other munitions ignited accidental fires, which burned into the surrounding forest. These mimicked the natural fires that had persisted for thousands of years and created better conditions for the woodpeckers to survive. Switching the burn schedule to the growing season began to restore open, savanna-like conditions across much larger sections of the base.

Biologists discovered, however, that the fires alone were not enough to quickly increase the RCW population. They needed another tool.

SETTING EGLIN AFLAME

"Here at Eglin our goal is to burn 90,000 acres [36,000 ha] a year to keep all our managed habitats on a maximum of three years between fires," explains Brett Williams, who helps manage the burn program on the base. "While most prescribed fire programs are burning only hundreds of acres at a time, our average burn is about two *thousand* acres [800 ha]."

Before each burn, Williams's team has to submit a detailed plan that explains where they're going to burn and why, and what kind of weather conditions they need for the burn. Eglin's fire team usually can manage the fires with a crew of between six and twenty people.

"We burn with torches, and we have some ATVs that have torches mounted on them, but it's more efficient to use a helicopter for larger burns," Williams says. "We use this machine mounted within the helicopter that shoots out little 'ping-pong balls' called DAIDs—Delayed Aerial Ignition Devices. These are loaded with a dry chemical called potassium permanganate. As the balls go down a chute attached to the machine, they get injected with ethyl glycol. Twenty or thirty seconds later, they catch fire and set the vegetation aflame. It takes a lot of training and coordination, but we can burn about 1,000 acres [400 ha] in an hour."

And what do the military brass—the high-ranking leaders—think of all this burning?

"They're 100 percent behind us to keep burning because if we don't and we get a big wildfire, it can shut down our military test and training missions," says Williams. "Then the military has to get out of our way so we can fight the fire, and it backs up training schedules. Or if they're testing some weapons system, a big fire interrupts what they're doing, costing them time and money."

BIRDS WITH CAVITIES

One of the obstacles to growing Eglin's RCW population, it turned out, was the shortage of nesting and roosting cavities. "It can take the birds quite a long time to drill out a new cavity," explains endangered species biologist Jeremy Preston. "It can vary from tree to tree and from bird to bird, but as we began actively managing and growing woodpecker populations, we found that the artificial cavity was a very necessary tool."

Eglin's biologists have two options for creating artificial cavities. One is to use a chain saw to carve a little cube from the tree and then pop a prebuilt wooden nest known as a box insert into the hole. Unfortunately, these "woodpecker condos" don't last long. The wood rots away in a few years, and the birds stop using them. Carving out a big chunk of wood also weakens a tree. When hurricanes roar through the area, trees often snap off right where woodpecker boxes have been inserted.

Because of these issues, Eglin's wildlife team usually drills artificial cavities directly into the trees. "We can get a new cavity for the birds real fast," explains Preston. "You pretty much drill two holes, one straight into the tree, and a second hole down at an angle that intersects the first hole and that you eventually seal off. This forms a little upside-down L- or 7-shaped cavity that mimics what the woodpecker's holes naturally look like."

Unlike many ordinary birdhouses, this artificial nest box is made to look as if it is part of the tree.

RED-COCKADED REBOUND

Since 1993, biologists have drilled more than fifteen hundred artificial cavities for RCWs. This housing boom, combined with improved burning techniques, quickly worked its magic at Eglin. The first reliable estimate of RCWs at Eglin, which occurred in 1992, was 235 active groups of breeding birds. Biologists refer to these as potential breeding groups, or PBGs. The following two years, the number of PBGs declined to a low of 217. From then on, the new RCW management activities allowed the woodpecker population to become dramatically larger.

Eglin and the USFWS agreed that when the base's RCW number reached 350 potential breeding groups, the population would be declared "recovered." Eglin exceeded that number in 2009, when they reached 371. Just recovering the RCW, however, did not meet Eglin's own goals. The base needed to be able to continue its ever-expanding military tests and exercises without worrying if one or two woodpeckers accidentally got killed. After some discussion, they reached an understanding with USFWS: if Eglin could increase the RCW population to 450 PBGs, the USFWS would not hold the base responsible for an occasional accident.

Eglin reached the 450 PBG target in 2016, and by 2023, it was home to more than 500 RCW potential breeding groups. Getting there has taken hard work. When biologists began their restoration efforts, some areas of the forest had too much undergrowth to be killed by fire alone. Eglin's resource managers have had to kill some of it with herbicide or cut it down and remove it manually. Drilling new woodpecker holes isn't always a picnic either.

"Sometimes you can tell that a tree with heart rot might actually be full of water," says biologist Kelly Jones, chuckling. "That's happened to most of our crew over the years. They've been strapped to a tree and drilled into it. When they pull out the drill, a column of water just 'fire hoses' them—usually in the winter. You're strapped on top of a ladder, and you have to wait for the whole column of water to empty out."

Despite the occasional dousing, the RCW conservation work at Eglin has been a resounding success. Not only has it helped the woodpeckers, but it also has allowed the military to continue its mission uninterrupted. Perhaps even more important, the changes spurred by RCW management have led to a vast improvement in the health of Eglin's entire longleaf pine ecosystem. This has paved the way for helping many more species—including a gray, domed reptile that looks like a miniature battle tank.

Male and female RCWs look almost identical except that males (*shown*) have tiny red patches on their heads.

One method people use to track the size of the RCW population is to catch individual birds, place aluminum or plastic bands around their legs, and release them. These bands can help identify each bird and provide information about how far it travels and how long it lives.

»» CHAPTER FOUR ««
ZOMBIE TORTOISES

The gopher tortoise earned its name for the deep burrows it digs—similar to a gopher.

Jeremy Preston kneels in the sandy soil and gently pushes a long cable into the mouth of a burrow. He's not sure how long the burrow is, but he has 25 feet (7.6 m) of cable to work with.

"Hopefully, we'll reach the end before the cable runs out," he says.

The cable has a tiny spotlight and a camera lens at its tip and carries a live video feed back to a monitor. As Preston pushes the cable forward, he glances at the monitor next to him to see inside the burrow.

"I'm hoping we might see a diamondback rattlesnake," he says, "or maybe a coachwhip or other snake."

Alas, no snake of any kind appears on the monitor, but the camera does pick up a cricket and some kind of worm.

Then a smooth gray shape appears. It fills most of the width of the burrow.

"There it is," Preston says, sitting back with a smile. "A gopher tortoise."

KEYSTONE REPTILE

Gopher tortoises are one of five tortoises native to North America and the only species living east of the Mississippi River. Like the red-cockaded woodpecker, the gopher tortoise is closely associated with longleaf pine forest. While it does also live in other forest types, its range almost exactly mirrors the original range of the longleaf pine. Gopher tortoises can be up to 15 inches (38 cm) long and weigh as much as a good-sized bowling ball—14 or 15 pounds (about 6 or 7 kg). As with most turtles, mortality for young animals is high. But once they reach a larger size, they are thought to have a life span of up to eighty or one hundred years.

The defining feature of the gopher tortoise is its ability to dig burrows. These burrows can reach 10 feet (3 m) deep and 25 to 35 feet (7.6 to 10.7 m) long. They serve as refuges for the tortoise to escape predators, wildfires, and extreme temperatures. In Florida the outdoor air temperature can exceed 100°F (38°C) or drop below freezing, but air inside a burrow typically hovers

Jeremy Preston uses a video camera attached to a long cable to observe the inhabitants of a gopher tortoise burrow.

The largest snake native to the United States, the endangered eastern indigo snake, makes its home in gopher tortoise burrows. Its population has declined along with the population of gopher tortoises.

between 68°F and 72°F (20°C to 22°C). Gopher tortoises spend about 80 percent of their time in their burrows. Especially during cold winter months, they can remain underground for days or weeks at a time.

The tortoises are not the only ones that benefit from their burrows. Approximately three hundred invertebrate and sixty vertebrate animal species have been observed using gopher tortoise holes. These include many kinds of insects as well as larger animals such as the Florida pine snake, eastern diamondback rattlesnake, gopher frog, Florida mouse, and eastern indigo snake. So many other animals use gopher tortoise burrows that scientists consider the tortoise, like the red-cockaded woodpecker, a keystone species.

Unfortunately, many of the same developments that impacted red-cockaded woodpeckers have caused gopher tortoise populations to decline too.

PLENTY OF PERILS

Many things have harmed gopher tortoises. The increase in roads and highways has been a major factor, as it has for a lot of wildlife. The tortoises are attracted to the open surroundings of roads but move across them so slowly that the animals are common casualties of automobiles. Studies have shown that when a new road is built, reptile populations in the area plummet.

"Another reason they declined is people ate them," Preston explains. "One gentleman I go to church with is ninety-three years old and has lived around here for a long time. He told me that he remembers going to the butcher years ago, and chicken was the cheapest thing for sale. Pork was the next cheapest and then beef and you go up the list until veal was the most expensive. Then he said that gopher tortoise meat was even *more* expensive than that. It was a delicacy."

Fire suppression has had one of the most drastic impacts on gopher tortoises. "Tortoises like really open-canopied forests," Preston says. "They're a reptile so they need lots and lots of sunlight for their bodies to stay warm and function. But they also are herbivores,

In the warmer southern regions of Florida, gopher tortoises can be active year-round. But in the northern parts of their range, they typically stay in their burrows during the winter months.

and they eat that grassy stuff that grows on the ground." Without frequent fires, an understory of shrubs moves in, blocking sunlight for low-growing grasses and other plants that tortoises depend on. Shrubs also make it harder for tortoises to warm their bodies to optimal temperatures. "Tortoises will leave forests that are getting too shrubby and too dark," Preston explains, "and they'll find road shoulders, which increases the risk of them getting run over."

By the 1980s, gopher tortoise populations had decreased enough that both state and federal governments began legally protecting the animal. The tortoise is currently protected as a threatened or endangered species in Louisiana, Mississippi, Georgia, Florida, and South Carolina, and is protected as a nongame (cannot be hunted or caught) species in Alabama. The population west of Florida declined so dramatically that in 1987, the USFWS listed it as threatened under the Endangered Species Act.

BASE TORTOISES

At Eglin, roads were not as much of a problem as in many other areas, but the tortoises were still struggling. Even though the red-cockaded woodpecker management program had returned fire to the landscape, surveys starting in 2010 showed that tortoise populations continued to decrease. In four out of five study sites, active tortoise burrow numbers dropped by about half—and sometimes more—over a period of several years. In any particular area, gopher tortoises need a population of about 250 animals to survive and keep reproducing, but no place on Eglin came close to having those numbers. As with the RCW, the largest surviving populations were near busy airfields and munitions testing ranges. There, accidental fires and human clearing of vegetation kept the landscape open.

"That's where all of our [tortoise] populations were," Preston recalls. "They were on our bombing ranges and our airfields, and in a few strips of power line right-of-ways." Right-of-ways are places where large power lines cut through an area and power companies have cleared out tall vegetation. "These tortoises," Preston continues, "migrated to these places slowly through time as the rest of their habitat was degraded. And even though we started a lot of restoration work and brought back prescribed fire to open areas back up, the tortoises became kind of trapped in those other places."

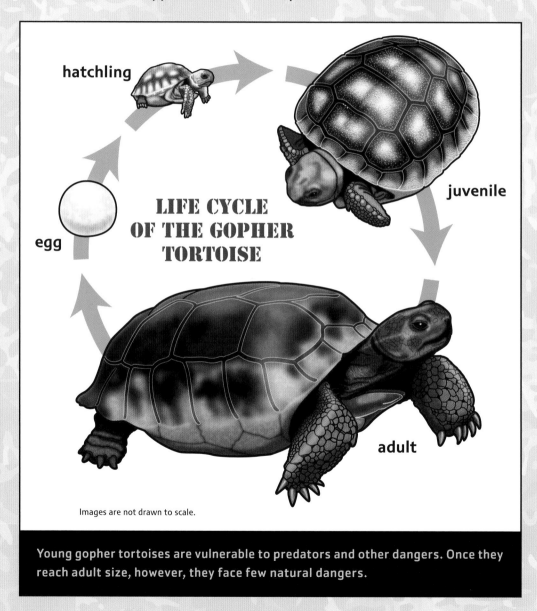

Young gopher tortoises are vulnerable to predators and other dangers. Once they reach adult size, however, they face few natural dangers.

And that left the base with a dilemma. By reintroducing fire onto the base, Eglin had a lot of perfect habitat for gopher tortoises. What it did *not* have were enough tortoises to repopulate those areas. Biologists were hesitant to try to move any of the adult tortoises from the firing ranges and airfields because they were surviving pretty well there. But the baby tortoises born there did not have the ecosystem structure they needed to survive. Young tortoises often prefer fallen trees and other objects to burrow under. Without that shelter, many are killed by predators, exposure to extreme temperatures, and accidents. So why not move the babies into the newly restored areas? Because biologists knew that even under the best conditions, only a small percentage of them would ever survive to adulthood, and those that did would take decades to begin reproducing.

Eglin's biologists didn't see how they would reintroduce tortoises into all the suitable habitats that were waiting for them. Then, almost out of nowhere, a huge supply of adult tortoises practically landed in their laps.

FORGOTTEN REPTILES

"One day, about 2014 or 2015," Preston recalls, "we got a phone call from Nokuse Plantation, a large private nature reserve to the east of us. They said, 'Hey, we need some help. There's a lot of tortoises over here that need a place to go and it's just costing us a lot of time and effort to receive and manage them.'"

Nokuse Plantation's animals came from a quirk in Florida's history of protecting tortoises. In Florida, one of the big threats to gopher tortoises is real estate development. Few places in the United States have seen such rapid growth of cities and towns, and much of this development has occurred in prime gopher tortoise habitat. In 1991, when concern for the species was growing, the state began requiring developers to buy permits for any gopher tortoises that might be killed during a construction project. These were called incidental take permits, and the money collected from the permits was supposed to be used to buy and protect other tortoise habitat. Unfortunately, the state spent very little of this money helping tortoises. And while developers were urged to relocate tortoises to other places, they were not *required* to do so. Under the system, construction activities continued to kill thousands of tortoises.

In 2007, when Florida listed the reptile as threatened under state law, it tightened up its tortoise regulations. After that, developers legally *had* to safely remove and relocate tortoises before any construction could begin. Places such as Nokuse Plantation that had extra tortoise habitat could charge fees to real estate developers to accept these rescued tortoises. The catch? When it introduced its stricter tortoise laws, the State of Florida decided to honor the old incidental take permits. If a developer had bought a permit years earlier, *it still could kill tortoises without any fines or penalties*.

A woman named Carissa Kent recognized this loophole. But she also realized that many developers with the old

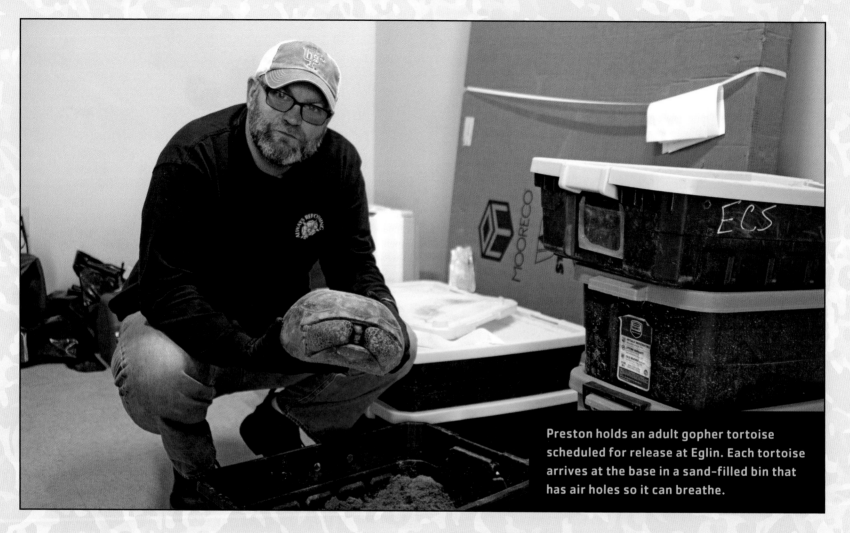

Preston holds an adult gopher tortoise scheduled for release at Eglin. Each tortoise arrives at the base in a sand-filled bin that has air holes so it can breathe.

permits had not yet started or finished construction on their properties. Kent created a nonprofit group called Saving Florida's Gopher Tortoises and began reaching out to these developers. She asked them if her group could rescue these tortoises and move them to safer locations. The problem with this plan was that the developers were not required to pay anyone to catch and move the animals. Nokuse Plantation and similar facilities also wouldn't receive any money for taking the tortoises, and that made it harder for them to continue funding their conservation activities.

"When we got the call from Nokuse," Preston recalls, "we at first said well, we don't want to take them if you're getting paid to receive these animals. You're doing great work, and the air force doesn't want to rob you of revenue generated from tortoise conservation." But the Nokuse Plantation assured him that they were only trying to relocate the tortoises that have no money associated with them.

COMING BACK TO LIFE

Once Eglin's biologists understood where these new tortoises were coming from, they jumped in to save them. "The Fish and Wildlife Service found out about it," says Preston, "and the Florida Fish and Wildlife Conservation Commission found out about it, and they were all encouraging Eglin to take the tortoises because we had all of this prime habitat. It was in great shape for tortoises, but we just didn't have any tortoises to fill it." Taking the tortoises would be a win-win for everybody. It would save thousands of tortoises doomed to die under the old incidental take permitting scheme. And it would help restore vital keystone animals to thousands of acres of Eglin's longleaf pine ecosystem where the animals would receive long-term protection from surrounding development.

Preston and other biologists determined that Eglin had eighteen areas that would provide excellent tortoise habitat. From October 2017 to July 2019, a total of 1,421 tortoises were placed in these locations. To complete the project, Eglin's Natural Resources Branch worked with scientists at Texas A&M University as well as representatives from Saving Florida's Gopher Tortoises.

Each gopher tortoise area contains enough suitable habitat to accommodate at least 250 tortoises—the ideal number to create a sustainable population. Biologists enclose a portion of these areas with soft netting called silt net to prevent the tortoises from wandering off.

This silt net surrounds a new area where gopher tortoises have been released at Eglin.

"They have a tendency to want to get back to where they came from," Preston explains. "It's like 'This ain't my home. Where is my burrow I've lived in for the last thirty years?' They still try to get back home so we have to put up some sort of temporary physical barricade to keep them from just wandering off to wherever they came from—which in some cases might be 300 miles [500 km] away." Fortunately, the tortoises accept their new homes after a few months. They establish territories and form relationships with other tortoises. "After about eighteen months," Preston continues, "we remove the silt net so the tortoises are free to move about as they see fit, but they typically don't expand out very far. They don't have the need to wander since the habitat of their new home is really great."

To help the tortoises get settled, biologists dig them starter burrows that are about 3 feet [0.9 m] deep. The animals can be picky about which ones they choose. "Eight out of ten times," Preston says, "they'll go down the burrow and come right back out and just start wandering around and say, 'I don't like this one.' We don't take it personally. Usually we'll come back the next day and some other tortoise has found what you dug and is now using it."

TORTOISES BY THE THOUSANDS

By 2024 more than ten thousand tortoises had been relocated to Eglin Air Force Base. Preston and other biologists call them zombie tortoises because many—perhaps most—of them would have been killed if they had not been moved to the safety of Eglin. Sort of like fictional zombies, these tortoises that were doomed to die have been given a new chance at life. They are reproducing and show every indication of creating long-term, healthy populations. Even better, Eglin has room to resettle thousands more.

As the tortoises return to Eglin's landscape, they help out other animals that use tortoise burrows. Scientists placed game cameras in or next to some tortoise burrows. The cameras captured images of several kinds of snakes and amphibians, as well as raccoons, foxes, bobcats, and coyotes using or hanging around the burrows. Some of these animals eat small gopher tortoises, but the large number of breeding tortoises at Eglin should ensure that plenty survive to adulthood.

"Tortoises play the long game," says Preston. "Only 2 or 3 percent of hatchlings survive to adulthood, but adults live eighty years, sometimes longer. They take ten to twenty years to reach sexual maturity. But if you lay eggs for sixty out of your eighty years on the earth and you're getting 3 percent to survive, you're definitely replacing yourself and your mate, with some extra left over. I'm really, really happy with how they're doing."

As gopher tortoises mature, they lose their bright yellow, orange, and brown coloration and turn a darker brown or gray.

#》》 CHAPTER FIVE 《《
SALAMANDER SOLUTIONS

Adult reticulated flatwoods salamanders' light markings can change in their lifetime, often starting out in a patchy pattern before morphing into stripes or bands. Some even turn fully black.

Eglin's commitment to restoring its longleaf pine ecosystems is seeing major payoffs for RCWs, gopher tortoises, and many other animal and plant species. But challenges remain to protect other species—including one of the base's smallest residents, an endangered amphibian known as the reticulated flatwoods salamander, or RFS.

The reticulated flatwoods salamander owes its name "reticulated" to the pale, tigerlike stripes running along its body. "Flatwoods" comes from the fact that it generally lives in areas with flat, sandy terrain. Despite its fancy name, the salamander doesn't make a big first impression. Adults easily fit in the palm of your hand and have drab, camouflage coloring that protects them from all but the sharpest eyes.

No one knows how many reticulated flatwoods salamanders remain, but it is probably no more than a few thousand. In 2009 the USFWS officially listed the reticulated flatwoods salamander as endangered. Although small populations live in southwest Georgia and one or two other spots, the animal's last great stronghold occurs in the pinewood wetlands of Eglin Air Force Base and the neighboring Escribano Point Wildlife Management Area. For the past ten years, biologists have been trying to increase this precarious population.

First, however, they have had to untangle its remarkably complicated life cycle.

DRYING UP

Kelly Jones is a private contractor working for Virginia Tech's Fish and Wildlife Conservation Department. He leads the field team working to save the RFS. Eglin and the DoD hired scientists from Virginia Tech mainly to help recover the base's red-cockaded woodpecker population. As the demands of woodpecker conservation eased up, however, Eglin turned to Virginia Tech to work on other conservation projects, particularly saving the endangered RFS.

"Our salamander work really started picking up in 2010," Jones recalls. "When I got here, I was completely ignorant of the animal's life history, but I came to realize there was still a lot that is unknown about the animal. So I started calling people who had studied the salamander back in the early 1990s. I asked them, 'When was the last time you saw the salamanders? Were they on this property? What about that property? Are you still finding them?'"

THE SALAMANDER SPLIT

The RFS used to be known as the flatwoods salamander. But in 2007 scientists determined that the flatwoods salamander should be split into two species, the frosted flatwoods salamander and the reticulated flatwoods salamander. The frosted salamander's range includes several counties in north-central Florida and isolated locations in Georgia and South Carolina. The reticulated flatwoods salamander's range extends from southwest Georgia and a thin slice of Alabama to parts of the western Florida Panhandle.

These conversations confirmed how little was known about the species. Jones also realized that even in the last two decades, the salamander had disappeared from many of its former locations. "Its population," he says, "was like an ocean drying down to little pools."

A COMPLICATED EXISTENCE

The first priority for the Virginia Tech team was to figure out where the salamanders still lived and why they survived there. Finding any sort of pattern proved difficult. Salamanders were absent from many natural-looking wetlands that seemed as though they should provide good habitat for the animals. And they popped up in places where it seemed they should have been killed off. These included construction zones and areas that had experienced ocean flooding from hurricanes. Confusing the issue even more was a poor understanding of what made good RFS breeding habitat.

"So the basic thing that you would hear [from other natural resource specialists]," Jones recalls, "is that flatwood salamanders breed on the edges of certain kinds of swamps with a ring of wiregrass growing around the edge. That led us to believe that we just needed to open up the edges of these wetlands so that a grassy ring can grow, and the salamanders can breed."

That turned out *not* to be the case.

Even before the Virginia Tech team began working with the salamanders, scientists working for the Florida Fish and Wildlife Conservation Commission had figured out that the animals primarily breed in *ephemeral wetlands*—wetlands that completely dry up during the dry season. Virginia Tech and the conservation commission scientists had also experimented with

Kelly Jones inspects new plant growth following a controlled burn in one of Eglin's ephemeral wetlands.

clearing some hardwood trees from certain wetlands, and they realized that this improved breeding for the salamanders. These discoveries and other research helped scientists recognize that the main reason salamanders were disappearing from Eglin was the same reason red-cockaded woodpeckers and gopher tortoises had disappeared: fire suppression. But for the salamanders, the need for fire has an extra twist. For the salamanders to thrive, fire can't just burn to the edges of wetlands. It has to regularly burn *all the way through the wetland basins*—the areas where water pools up for at least part of the year.

The reasons for this are complicated. Old wetlands that haven't burned in a long time become shaded and choked out by trees and other woody vegetation. Such wetlands do not provide enough food for the salamanders. They also don't produce the types of low-growing plants the salamanders need to lay eggs in or for their young to hide in. A good burn through the heart of a wetland, though, serves as a "hard reset" of the habitat. The fire leaves behind open, sunny conditions just right for the salamanders.

"So what we had going for us on Eglin," Jones explains, "is that because of the woodpeckers, we already had a good burn program. The drier upland areas were in decent shape. So were the transitional zones between these drier uplands and the wetlands. But it was actually the wetland basins that were most affected by fire suppression. While the rest of the landscape had been managed for fire, these basins had been locked in a time capsule of fire suppression."

The interlocking leaves of tiny ground plants provide ideal hiding and nesting places for reticulated flatwoods salamanders.

THE RISKS OF EPHEMERAL WETLANDS

If you're a flatwoods salamander, breeding in ephemeral wetlands requires perfect timing—and a lot of luck. In the fall and early winter, the salamanders move into ephemeral wetlands to find a mate and lay eggs. Most other closely related salamander species lay their eggs in water, but flatwoods salamanders deposit their eggs on dry land. The RFS embryos develop inside the eggs but are fully aquatic larvae (tadpoles) with gills for underwater breathing. That means that they need water to hatch out into. And *that* means that rains must fill the ephemeral wetlands at just the right time. If the salamanders lay eggs during a drought year, or they lay them too high up on the shore where the water doesn't reach them, the larva in the eggs will die from drying out or use up the energy in their yolks and starve to death.

Still, if the weather cooperates, this risky breeding strategy has some advantages. One is that ephemeral wetlands tend to have fewer fish and other large aquatic predators than year-round ponds do. If rains *do* come at the right time with enough water, tadpoles have a better chance to survive. Developing early also gives the tadpoles a significant advantage over other pond species. That's because species that lay their eggs in water have to wait until the wetlands start to fill before their eggs can even start developing.

"By laying its eggs on land," Jones explains, "the RFS gets a head start on the whole process. Their larvae are already developing inside their eggs before the water comes. This sets up the tadpoles to be dominant, significant predators in this small-scale ecosystem."

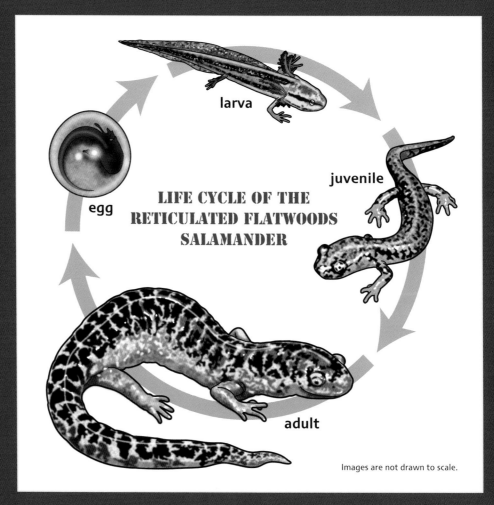

LIFE CYCLE OF THE RETICULATED FLATWOODS SALAMANDER

Images are not drawn to scale.

The life cycle of the RFS is similar to that of other amphibians—but it requires rains to arrive at just the right time to succeed.

CHURNIN' AND BURNIN'

Once biologists realized how important it was to burn the heart of the wetlands, they launched an extensive program to restore habitat that would support the RFS. This was harder than it sounds. After years without fire, many of these wetland basins were filled with large trees and other growth that could not be killed by fire alone. Some of the basins also didn't dry out enough to burn every year. If water remained standing in a basin throughout a dry season, the team would have to wait until a drier year to conduct the burn.

The salamander's population status and biology also complicated the situation. Salamander populations were so small that even if scientists created a ton of new habitat, it might take years for the animals to recolonize it—and weather would play a key role. Rains had to come at just the right time and in the right amounts for the eggs to hatch and for the young salamanders to survive. Unfortunately, the South has experienced frequent droughts in recent years, killing eggs and tadpoles. In some years, female salamanders may not have had enough food to attempt breeding.

Nonetheless, one wetland at a time, the salamander team began clearing and burning wetlands to open them up for salamanders. Often they hired workers to cut down trees by hand and then burn them. By the end of 2022, the Virginia Tech team had restored more than 185 acres (75 ha) of habitat covering fifty wetlands. It also released salamander larvae from established populations into two newly restored wetlands where salamander populations had "blinked out" fifteen years before.

Was the project working?

This is one of several Eglin wetlands that has been restored to support healthy RFS populations.

CREEPING FORWARD

Adult salamanders are so difficult to locate that Jones is reluctant to say how well the team's efforts are paying off. He believes that the entire salamander population at Eglin may number only one or two thousand animals—maybe fewer. Still, his sense is that their situation is improving.

"The females are picky about the conditions they need," he says, "so they're kind of a Goldilocks species. But with that said, we've been able to observe new sites that have salamanders in small numbers." Evaluating progress has been difficult because the salamanders have only ever been documented at twenty-nine sites on Eglin, and when Jones and his team started their work, the animals had disappeared from most of those sites. "But we have seen a couple of new sites with salamanders," Jones says, "and other sites that have had salamanders return after years and years of not finding any salamanders."

Put it all together and Jones believes that with a little help from the weather, he and his colleagues are creating the right conditions for success. "If we can get them past that tipping point," he says, "where they can really start getting some good years, I think they're capable of multiplying in large numbers."

That could lead to everyone's ultimate goals: stabilizing the salamander's population and getting them removed from the endangered species list.

EGLIN AT THE BEACH

Although Eglin's large mainland area is better known, the base also controls some of the most pristine coastline in the Florida Panhandle. The coastline is on two of Florida's barrier islands. "Technically we have 20 miles [32 km]," explains Christina Mahmood, Eglin's Natural Resources Branch office volunteer coordinator. "Three of those miles [5 km] are a few hours to the east of us at Cape San Blas, and we have about 17 miles [27 km] here on Santa Rosa and Okaloosa Islands." About 4 miles (6.4 km) of the coastline are open to the public. The rest is reserved for military operations. This makes it an invaluable refuge for coastal plants and animals.

One of Mahmood's main jobs is to organize volunteers to conduct sea turtle surveys. "Four different sea turtle species nest in the Florida

Christina Mahmood watches as a newly hatched sea turtle makes its way to the water.

Panhandle," she says. "Loggerheads and green sea turtles are our primary nesters, followed by Kemp's ridley and leatherbacks."

All sea turtles are federally protected in the United States, and several aspects of Eglin's beaches make them especially attractive for nesting. One is the lack of lights. After they emerge from the nest, hatchling sea turtles are drawn to moonlight over the water. Artificial lights from hotels, restaurants, and other developments often confuse the young turtles. Many end up on roads and parking lots where they get run over or die from dehydration—something that rarely happens on Eglin property. The lack of public access at most of Eglin's coastline also makes sea turtle nests less likely to be trampled there than on public beaches. Of course, beach landings and other military operations aren't great for the turtles.

"Since the military utilizes the beach during nesting season, there's a lot of times we have to come through and survey ahead of a mission," Mahmood explains. "The military has to wait until we're done, and there are two areas on the beach that are considered relocation zones. Anytime we find a nest in one of those two areas, we have a permit from the Florida Fish & Wildlife Conservation Commission to carefully move that nest to a different location. We've all gotten really good at digging nest holes that resemble flipper holes dug by a female turtle," she says with a laugh.

Other vulnerable species that depend on Eglin's barrier islands include the Santa Rosa beach mouse, the Florida perforate lichen, and a variety of shorebirds including red knots, piping plovers, and least terns. Mahmood also organizes volunteers to monitor shorebirds, including state-threatened snowy plovers that nest behind the islands' first row of sand dunes. These plovers produce an average of sixty nests at Eglin each year, and the base's isolation helps protect these birds from human activities. "Their eggs are really well camouflaged," Mahmood says, "and they nest on the ground so it's really hard to see them. But if people were allowed behind the dunes, I imagine nests would get stepped on."

Eglin's protected beach and dune areas provide essential refuges for piping plovers and many other species.

»»» CHAPTER SIX «««
THE BIGGER MISSION

The desert habitat at Fort Irwin, California, is home to several endangered species including the desert tortoise and Lane Mountain milk vetch.

Eglin Air Force Base is not alone in wrestling with the challenges of protecting America's plant and animal species. According to the DoD Natural Resources office, 341 military bases covering 25 million acres (10 million ha) of land contain significant natural resources that require active management. Large bases such as Camp Pendleton in California and Homestead Air Reserve Base in Florida each protect about two dozen threatened and endangered species. They also protect unique ecosystems that are being gobbled up by surrounding cities.

Smaller bases also are vitally important. Six army and navy bases in Hawaii and Guam each hold between twenty-six and fifty-eight threatened and endangered species, including many birds that are close to extinction. And although the DoD originally began protecting species because it was forced to by law, most modern military leaders view conservation as a part of their larger military mission.

Brigadier General Jeff Geraghty assumed command of Eglin in 2022 and has a biology degree from the United States Air Force Academy. He explains:

During his service in the US Air Force, General White helped kick-start many environmental projects. Named in his honor, the Thomas D. White Environmental Awards are granted each year to the air force installations, teams, and individuals that implement the best environmental projects.

> We have here a patriotic sense that is part of the defense job. In fact, our fourth Chief of Staff for the United States Air Force, a guy by the name of General Thomas White, said, "The mission of the Department of Defense is more than just aircraft, guns, and missiles. Part of the defense job is protecting the lands, waters, timber, and wildlife—the priceless natural resources that make this great nation of ours worth defending." So I know this team is dedicated not just in their sworn oath to protect and defend the Constitution but to the diversity of our natural resources and ecosystems too.

CONSERVATION ACROSS THE NATION

Like Eglin, the army's Fort Bragg in North Carolina has extensive stands of longleaf pine and a large population of red-cockaded woodpeckers. There, biologists have performed much of the same kind of work as Eglin to protect both species. Accidental fires from artillery explosions have also maintained critical habitat for the Saint Francis' satyr, a highly endangered butterfly that lives only in the Fort Bragg area. Military and university biologists are working together to ensure the butterfly's survival and create additional habitat for the insect. Bases in other states and territories face similar—and far different—conservation challenges.

The Saint Francis' satyr butterfly lives in open grassy wetlands that are maintained by fires and beavers.

MARIANA ISLANDS & GUAM

The Mariana Islands, which include the island of Guam, is a vital hub for DoD operations. Millions of tons of commercial and military cargo come in and out of the islands—cargo that, unfortunately, has carried devastating invasive species. These include the brown tree snake, coconut rhinoceros beetle, and little fire ant. The islands' native wildlife evolved without any defense to many invaders. The brown tree snake alone has eliminated most of Guam's native birds, along with bat and lizard species. Through its biosecurity program, the navy has implemented strict measures to inspect all military cargo. This has helped prevent the spread of the brown tree snake to Hawaii and other locations—and helped protect the Mariana Islands from new invaders.

The brown tree snake is also called the brown cat snake for the catlike appearance of its eyes.

HAWAII

Hawaii is home to almost five hundred federally listed threatened and endangered species—more than any other place in the United States. Clearing land for plantations, pastures, cities, and military bases has forced native species into smaller and smaller areas. Much like the Mariana Islands, Hawaii also has been hammered by invasive species. Biologists at Hawaii's military bases conduct a host of activities to protect the islands' native plants and animals. The army's Pohakuloa Training Area, for example, is home to a number of unique shrublands, forests, and grasslands. To safeguard these areas, the army has removed invasive goats and sheep that had been causing extensive damage by eating native plants. Working with other partners, they also constructed 86 miles (138 km) of fencing to keep these and other animals from reinvading these areas. Fires, fueled by invasive grasses, also threaten native ecosystems. Fire is an essential part of the natural cycle of longleaf pine forests, but that's not the case in Hawaii where most ecosystems evolved without fire. To help firefighters respond quickly to fires and limit their spread, the army has built a system of firebreaks. This system has already stopped one fire that threatened to destroy the entire remaining population of a native plant called the Mauna Kea pamakani.

CALIFORNIA

California has more than thirty major military bases representing every branch of the DoD, and many have extensive conservation programs. At Beale Air Force Base, biologists have teamed up with the USFWS, California Department of Fish and Wildlife, and other agencies to remove a dam on Dry Creek that prevented Chinook salmon and Central Valley steelhead trout from traveling upstream to spawn. They also worked to restore the riverbank, improving conditions for these and other native fish.

Farther south, in heavily populated Los Angeles County, the military has helped save an animal once thought to be extinct. The Palos Verdes blue butterfly lived only on the Palos Verdes Peninsula, and it was listed as endangered in 1980. But in 1983 the City of Palos Verdes knowingly bulldozed the last place where the butterfly lived, and it was thought to have disappeared forever. Remarkably, in 1994 biologists discovered sixty-nine of the butterflies at a military fuel depot called Defense Fuel Support Point. The Defense Logistics Agency, an arm of the military that provides supplies to all branches of the DoD, ran the fuel depot. As soon as the butterfly was rediscovered, the agency, US Navy, and Department of Defense began working with local partners to increase its tiny population.

The butterfly depends on two host plants, Santa Barbara milk vetch and deerweed. The DoD helped fund and establish a plant nursery at the fuel depot so that good butterfly habitat could be planted at other locations on the Palos Verdes Peninsula. It also provided funding to raise the butterflies by hand. Since the program started, volunteers have raised thousands of the butterflies, and the insects have been reintroduced to two other sites outside of the fuel depot. Slowly, the butterfly's population has increased to about three hundred, improving the chances that it will survive long into the future.

A Palos Verdes blue butterfly lands on a blooming deerweed plant.

Eglin's commitment to conservation is reflected in the number of awards it has received. The base has been selected as the Department of Defense's top natural resources program five times in recent years. It has also won many other awards for its environmental work. That doesn't mean that progress has been easy. As the world becomes more complicated, so does the job of protecting nature. Managing the fire program at Eglin, for instance, requires extensive coordination between biologists, fire operations experts, and those in charge of military testing and training exercises.

"It's difficult," forestry supervisor Al Sutsko admits. "Eglin air space, first of all, is super busy. You constantly have flights. The F-35 training school is here and they're flying every day." The army's 7th Special Forces Group, part of the Green Berets, also uses the base. "There are test areas all over the place," Sutsko continues, "and in a lot of those test areas, they all run different missions. Some are smoke sensitive so you can't dump smoke on them. [Mission planners are] not going to be happy if they're running a $100 million mission, and it gets scrubbed because of a prescribed burn!"

Besides managing prescribed burns, controlling harmful plants and animals is a giant job almost everywhere. This is especially true on Pacific Islands where invasive plants, reptiles, insects, and other species have taken huge tolls on native animals and ecosystems (see "Conservation across the Nation"). Eglin is a bit more fortunate than other places because its sandy soils don't support as many invasive plants. Introduced hogs, however, do extensive damage on the base. They tear up wetlands and other fragile places and have to be trapped and kept out of many areas.

LIMITED RESOURCES, BIG REWARDS

All of these conservation efforts require time and money—which can often be in short supply. So far, Eglin has received sufficient funds from the DoD to support its activities helping the red-cockaded woodpecker, reticulated flatwoods salamander, and other species. That may not be the case going forward.

"Although we would love to have more, we expect to have to continue with what we have," General Geraghty says. "One of the messages we hear constantly . . . is 'Don't come asking for more people and more money because we are constrained, and we have tight priorities.' So we will probably have to continue to figure out how to do more with less, or decide what we should stop doing."

Still, money spent on conservation is an essential investment, both for the military and the nation. The DoD has huge incentives to protect vulnerable species. If an endangered plant or animal could be harmed by military operations, that could interfere with vital training or testing programs. By making sure that threatened and endangered species are safe from extinction, the military mission can continue uninterrupted. For that to happen, however, the military can't just react to urgent conservation concerns. It has to anticipate the future and the challenges it will bring.

"We don't want to get caught where there's an endangered species that's in an emergency situation," Geraghty says. "That's why we monitor all of the biodiversity here on the Eglin installation. We're keeping an eye on the trends so that we don't have to take emergency measures that would really restrict the military mission. Continuing to plan and think ahead, I think, is going to be our biggest strength."

This forward thinking represents a huge change for military leadership. Not long ago, protecting plants and animals was viewed mainly as a burden and inconvenience. Today, thousands of people within the Department of Defense not only recognize but appreciate the tremendous biological and ecological treasures found on DoD lands. As climate change, urban development, invasive species, and other environmental problems impact our planet, military bases will only become more important to protecting the plants and animals that help make America a special place. That gives added urgency to funding and supporting the military's conservation mission. It also inspires all of us to recognize and help protect the amazing biological diversity that exists around us, no matter where we live.

When threatened and endangered species are protected, military operations on US soil can continue without delay.

AUTHOR'S NOTE

Defending Nature could not have been written without the cooperation of the United States Air Force and the help of many, many individuals. The quotations, and much of the additional information, in *Defending Nature* come from interviews and other direct communications with scientists and other experts. As indicated in the introduction, no records exist of the early red-cockaded woodpecker conversations between Eglin's base commander, air force lawyers, and the base's chief of the Natural Resources Branch. In this instance, I re-created the dialogue based on information gathered from my interviews and other research. I chose to do this to give a better, more lively sense of how events unfolded.

I would like to especially thank the following DoD employees and private contractors working at Eglin for giving their valuable time to teach me about conservation at one of our nation's largest and most important military bases:

- Brigadier General Jeff Geraghty, commander, 96th Test Wing, Air Force Materiel Command
- Bruce Hagedorn, chief of Natural Resources Branch
- Justin Johnson, supervisory wildlife biologist
- Kelly Jones, field coordinator, Virginia Tech
- Christina Mahmood, volunteer coordinator
- Joseph Meyer, lead archaeologist, Jacobs Solutions
- Jeremy Preston, endangered species biologist
- Michael Spaits, environmental public affairs
- Al Sutsko, chief of forest management
- Brett Williams, wildland support module lead

My additional thanks to Matthew Vandersande of the USDA's Natural Resources Conservation Service for helping me obtain an accurate map of longleaf pine distribution, Jeffrey Walters of Virginia Tech for answering questions about the history of RCW recovery at Eglin, and USAF technical sergeants Ashley Nicole Taylor and Andrew A. Davis for helping secure the permissions I needed to move forward with this project.

I owe huge thanks to Carol Hinz at Millbrook Press for believing in this project; my editor, Jesseca Fusco, for painstakingly working with me to improve the manuscript; and my agent, Karen Grencik at Red Fox Literary, for her perseverance in placing it. Finally, special thanks and hugs to Suzanne Spencer and Jim McClellan—longleaf pine stewards and the best ground crew an author could ask for!

GLOSSARY

ATVs: all-terrain vehicles; off-road vehicles used in rough terrain

biodiversity: the variety of species and ecosystems in an area

cavity: a tree hole that is used by birds and other animals for nesting, shelter, or other purposes

Department of Defense (DoD): the primary government agency that oversees all US military operations

disperse: to spread out, especially after hatching or being born

endangered species: plants, animals, and other living things that are in danger of becoming extinct

Endangered Species Act: a 1973 law requiring the US to protect threatened and endangered species and devise plans to recover them

ephemeral wetland: one that is wet only seasonally, such as during the rainy season

fire suppression: a policy of putting out and preventing wildfires

herbicide: a chemical that kills plants

invertebrate: an animal that does not have a backbone

jeopardy biological opinion: a legal document that is a warning for an agency or individual that its planned action may have legal consequences

keystone species: a species that many other species depend on and without which an entire ecosystem would deteriorate

larva: an immature stage, such as a tadpole or caterpillar, of an animal that has not yet become an adult

mandate: an order

munitions: military equipment, weapons, or ammunition

organic material: leaves, twigs, branches, and other material produced by living things

petrochemicals: chemicals created from petroleum (oil), coal, and natural gas

rehabilitate: to make something healthy or whole again after it has been damaged or harmed in some way

reticulated: having a netlike appearance or pattern

roosting: for a bird, resting or sleeping

Sikes Act: a 1960 law requiring military bases to provide recreational opportunities to the public whenever possible, and to protect, conserve, and restore the bases' natural resources

solvent: a chemical used for dissolving other materials such as paint

stealth technology: for an aircraft, technology or construction techniques that make it invisible to radar

threatened species: plants, animals, and other living things that may soon become endangered if their situations don't change

turpentine: a fluid obtained from living trees that is used mainly as a solvent and to make various chemicals

urban sprawl: the "spreading out" of cities and towns as they grow, swallowing up natural and open lands

vertebrate: an animal with a backbone

SOURCE NOTES

7 Justin Johnson, interview with author, October 19, 2022.

11 "Eglin Air Force Base History," Eglin Air Force Base, accessed November 16, 2023, https://www.eglin.af.mil/About-Us/Fact-Sheets/Display/Article/390964/eglin-air-force-base-history/.

13 Al Sutsko, interview with author, October 19, 2022.

14 Brett Williams, interview with author, October 19, 2022.

21 Bruce Hagedorn, interview with author, October 19, 2022.

22 Williams, interview.

22 Williams.

22 Williams.

23 Jeremy Preston, interview with author, October 19, 2022.

23 Preston.

24 Kelly Jones, interview with author, October 21, 2022.

27 Preston, interview.

27 Preston.

27 Preston.

28 Preston.

28–29 Preston.

30 Preston.

31 Preston.

32 Preston.

33 Preston.

34 Preston.

34 Preston.

34 Preston.

37 Jones, interview.

38 Jones.

38 Jones.

39 Jones.

40 Jones.

41 Jones.

42 Jones.

42 Jones.

42 Christina Mahmood, interview with author, October 19, 2022.

42–43 Mahmood.

43 Mahmood.

43 Mahmood.

45 Jeff Geraghty, interview with author, October 6, 2022; "General Thomas D. White," National Military Fish and Wildlife Association, accessed November 17, 2023, https://www.nmfwa.org/hof/general-thomas-d.-white.

48 Sutsko, interview.

48 Geraghty, interview.

49 Geraghty.

FOR FURTHER INVESTIGATION

Despite how important military bases are to protecting plants and animals, very little has been written about the military's conservation efforts—and almost nothing for young readers. Several children's books talk about gopher tortoises, red-cockaded woodpeckers, and other species including these:

Collard, Sneed B., III. *Woodpeckers: Drilling Holes & Bagging Bugs*. Missoula, MT: Bucking Horse Books, 2018.

Dunphy, Madeleine. *At Home with the Gopher Tortoise: The Story of an Endangered Species*. Berkeley, CA: Web of Life Children's Books, 2010.

Swinburne, Stephen R. *The Sea Turtle Scientist*. Boston: Houghton Mifflin Harcourt Children's Books, 2014.

For those wishing to dig more deeply, I especially recommend the following adult books:

Conner, Richard N., D. Craig Rudolph, and Jeffrey R. Walters. *The Red-Cockaded Woodpecker: Surviving in a Fire-Maintained Ecosystem*. Austin: University of Texas Press, 2001.

Dunleavy, Laura, Drue DeBerry, and David Pashley. *Pine Ecosystem Conservation Handbook for the Gopher Tortoise in Florida*. Washington, DC: American Forest Foundation, 2008.

Havlick, David G. *Bombs Away: Militarization, Conservation, and Ecological Restoration*. Chicago: University of Chicago Press, 2018.

Beyond these resources, web searches will reveal a variety of useful information and news about conservation happening at different locations. Some sample searches might be "Conservation Eglin Air Force Base" or "Endangered honeycreepers Hawaii." A search for "Fort Bragg Butterflies," for instance, may call up a fascinating podcast about how, like the red-cockaded woodpecker, an endangered butterfly has been surviving on bombing ranges on the base. Keep in mind that the most reliable and accurate sites are those operated by government agencies or other scientific sources, and beware of personal blogs or those with personal or political agendas. Your parents, guardians, teachers, and librarians can help provide you with guidance on performing safe, reliable searches.

SELECTED BIBLIOGRAPHY

Blanc, Lori A., and Jeffrey R. Walters. "Endangered Species Management and Monitoring on Eglin Air Force Base, Florida. 2021 Project Annual Report: Red-Cockaded Woodpecker Component." Dept. of Biological Sciences, Virginia Polytechnic Institute and State University, Blacksburg, VA, 2022.

Connolly, Patrick. "Saving Florida's Gopher Tortoises: Group Rescues Reptiles from Death by Development." Phys.org, July 5, 2022. Originally printed in the *Orlando Sentinel*. https://phys.org/news/2022-07-florida-gopher-tortoises-group-reptiles.html.

Dunleavy, Laura, Drue DeBerry, and David Pashley. *Pine Ecosystem Conservation Handbook for the Gopher Tortoise in Florida*. Washington, DC: American Forest Foundation, 2008.

"Eglin Gopher Tortoise Monitoring and Management: Executive Summary." Internal document, author unknown, obtained from Jackson Guard Natural Resources Office at Eglin Air Force Base.

"General Thomas D. White." National Military Fish & Wildlife Association. Accessed November 17, 2023. https://www.nmfwa.org/hof/general-thomas-d.-white.

"Gopher Tortoise Commensals." Florida Fish and Wildlife Conservation Commission. Accessed November 21, 2023. https://myfwc.com/wildlifehabitats/wildlife/gopher-tortoise/commensals/.

Hardesty, J. L., R. J. Smith, C. J. Petrick, B. W. Hagedorn, and H. F. Percival. "Status and Distribution of the Fourth Largest Population of Red-Cockaded Woodpeckers: Preliminary Results from Eglin AFB, Florida." Conference paper. Red-Cockaded Woodpecker Symposium III: Recovery, Ecology and Management, January 24–28, 1993, North Charleston, South Carolina. Also in David Kulhavy, Robert G. Hooper, and Ralph Costa, eds. *Red-Cockaded Woodpecker: Recovery, Ecology and Management*. Nacogdoches, TX: Center for Applied Studies in Forestry, College of Forestry, Stephen F. Austin State University, 494–502. https://www.frames.gov/catalog/40343.

Jackson, Clay. "Lost and Found: Rediscovering the Palos Verdes Blue Butterfly." Terranea Life. Accessed November 21, 2023. https://terranealife.com/rediscovering-palos-verdes-blue-butterfly.

"96th Test Wing." Eglin Air Force Base. Accessed November 16, 2023. https://www.eglin.af.mil/About-Us/Fact-Sheets/Display/Article/390959/96th-test-wing/.

Orndorff, Ryan. "Threatened, Endangered, and At-Risk Species on DOD Lands." Department of Defense Natural Resources. Accessed November 16, 2023. https://denix.osd.mil/nr/focus-areas/biodiversity/threatened-endangered-and-at-risk-species/resources-of-interest/terp-lands-fact-sheet/.

Petrick, C. J., and B. W. Hagedorn. "Population Status and Trend of Red-Cockaded Woodpeckers on Eglin Air Force Base, Florida." In Ralph Costa and Susan J. Daniels, eds. *Proceedings of the 4th Red-Cockaded Woodpecker Symposium: Road to Recovery*. Blaine, WA: Hancock House, 2004, 203–214. https://www.frames.gov/catalog/42846.

"Saint Francis' Satyr Butterfly." United States Fish and Wildlife Service. Accessed November 21, 2023. https://www.fws.gov/species/saint-francis-satyr-butterfly-neonympha-mitchellii-francisci.

"Sikes Act." United States Fish and Wildlife Service. Accessed November 21, 2023. https://www.fws.gov/law/sikes-act.

Smith, Garrett C., Mark W. Patterson, and Harold R. Trendell. "The Demise of the Longleaf-Pine Ecosystem." *Southeastern Geographer* 40, no. 1 (May 2000): 75–92.

"Summary of the Endangered Species Act." Environmental Protection Agency. Accessed October 23, 2023. https://www.epa.gov/laws-regulations/summary-endangered-species-act.

"U.S. Air Force Integrated Natural Resources Management Plan, Eglin Air Force Base, Florida, 2022." Internal document, author unknown, obtained from Jackson Guard Natural Resources Office at Eglin Air Force Base.

US Fish and Wildlife Service. "Recovery Plan for the Reticulated Flatwoods Salamander (*Ambystoma bishopi*)." South Atlantic–Gulf Regional Office, Atlanta, Georgia, 2021. https://ecos.fws.gov/docs/recovery_plan/20210615_RETICULATED_Flatwoods_Salamander_RP.pdf.

Wills, Matthew. "Road Density Threatens Turtle Populations." JSTOR Daily, June 26, 2021. https://daily.jstor.org/road-density-threatens-turtle-populations/.

INDEX

Eglin Air Force Base, 8, 10–11
   Cultural Resources Management, 16
   land restoration, 22, 24, 37–39, 41–42
   threatened and endangered species, 4–5, 7, 9, 42–43, 45, 48–49
     See also gopher tortoise; longleaf pine; red-cockaded woodpecker (RCW); reticulated flatwoods salamander (RFS)
Endangered Species Act, 8–9, 19, 29
ephemeral wetlands, 38, 40

fire management, 21, 24, 30–31, 41, 48
   benefits of, 22, 39
   fire prevention, 16, 28–29, 39, 47
frosted flatwoods salamander, 37
F-35 Lightning II, 12, 48

Geraghty, Jeff, 45, 48–49
gopher tortoise, 26–35, 37, 39
   appearance, 27, 35
   life cycle, 30

green sea turtle, 43

Hagedorn, Bruce, 21

incidental take permit, 31, 33

jeopardy biological opinion, 5, 9, 21
Johnson, Justin, 7

Kemp's ridley sea turtle, 43
keystone species, 19, 27–28, 33

least tern, 43
leatherback sea turtle, 43
loggerhead sea turtle, 43
longleaf pine, 11–16, 18–20, 24, 27, 33, 37, 46–47

Mahmood, Christina, 42–43

Palos Verdes blue butterfly, 47
piping plover, 43
Preston, Jeremy, 23, 27–34

red-cockaded woodpecker (RCW), 4–5, 9, 16, 18–25, 27–28, 30, 37, 39, 46, 48
   appearance, 24
   life cycle, 19
   potential breeding group, 24
red heart rot, 20
reticulated flatwoods salamander (RFS), 36–42, 48
   appearance, 36–37
   life cycle, 40

Saint Francis' satyr butterfly, 46
Santa Rosa beach mouse, 43
Sikes, Robert, 8–9
Sikes Act, 8–9
snowy plover, 43
Sutsko, Alexander (Al), 13, 48

White, Thomas D., 45
Williams, Brett, 14, 22
wiregrass, 21

PHOTO ACKNOWLEDGMENTS

Image credits: U.S. Air Force, pp. 4, 15, 45, 49; Rolf Nussbaumer Photography/Alamy, p. 5; Bill_Dally/Getty Images, p. 6; US Fish and Wildlife Service, p. 8; AP Photo/CQ Archive, p. 9; AllenJMSmith/Getty Images, p. 10; US Air Force/Kristin Stewart, p. 12; Clint Farlinger/Alamy, p. 13 (top); Heritage Art/Heritage Images/Getty Images, p. 14; Liu Guanguan/China News Service/Getty Images, p. 17; William Leaman/Alamy, p. 18; blickwinkel/Alamy, p. 20; Sneed B. Collard III, pp. 21, 27, 29, 32, 33, 38, 39, 41, 42; Michael Greenfelder/Alamy, p. 23; William Leaman/Alamy, pp. 24, 25, 56; George Grall/Alamy, p. 26; Danita Delimont/Alamy, p. 28; Peter Llewellyn RF/Alamy, p. 35; Dante Fenolio/Science Source, p. 36; Johann Schumacher/Alamy, p. 43; US Marine Corps/Lance Cpl. Alex Devereux, p. 44; Dave Pavlik/Fort Liberty Garrison Public Affairs Office, p. 46 (left); Narelle Power/Alamy, p. 46 (right); Natural History Collection/Alamy, p. 47; Digital Storm/Shutterstock, p. 53; imageBROKER/Alamy, p. 55. Backgrounds: freeart/Shutterstock (military abstract); Green angel/Shutterstock (black camouflage); Guki/Shutterstock (green moss); akram ahmed karam/Shutterstock (military texture); Kirill Mlayshev/Shutterstock (chevron).

Front cover: Dante Fenolio/Science Source (salamander); cturtletrax/iStock/Getty Images (turtle); Danita Delimont/Alamy (bird); R. Nial Bradshaw/US Air Force (planes). Back cover: imageBROKER.com/GmbH & Co. KG/Alamy (turtle); Digital Storm/Shutterstock (plane).